THE ZOO

written and photographed
by
Mia Coulton

I am at the zoo.

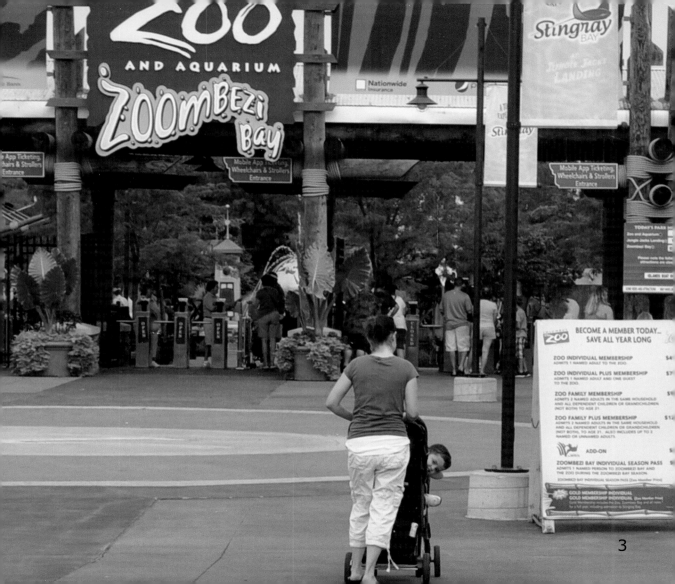

I see a tiger at the zoo.

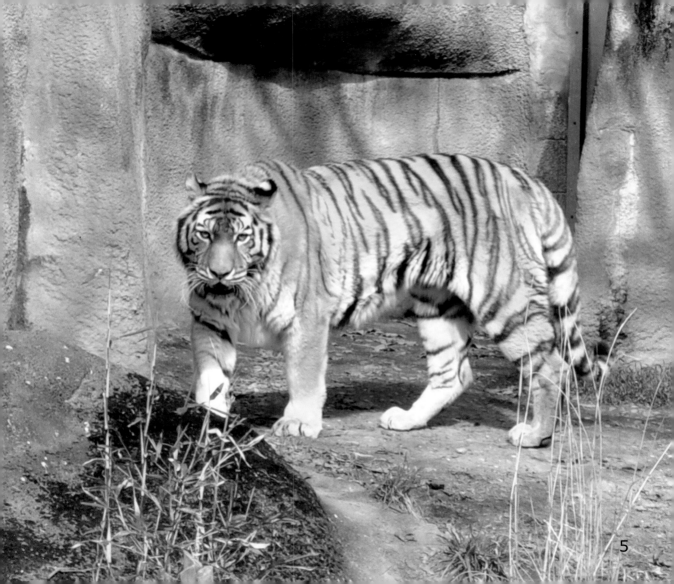

I see a bear at the zoo.

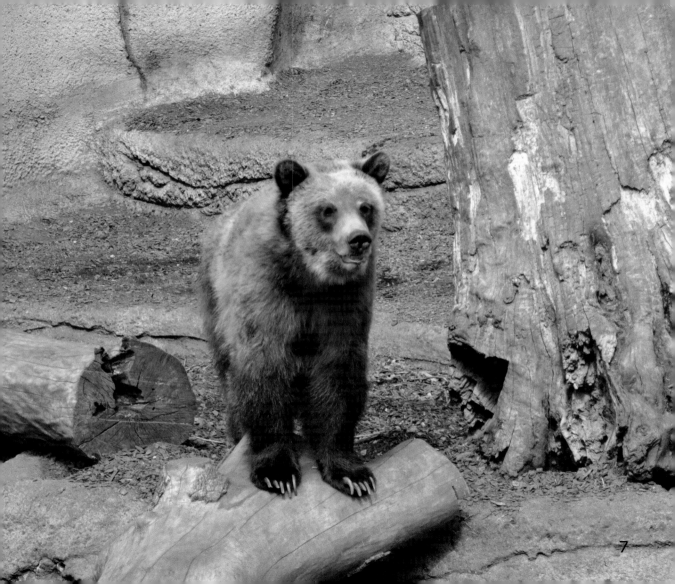

I see giraffes at the zoo.

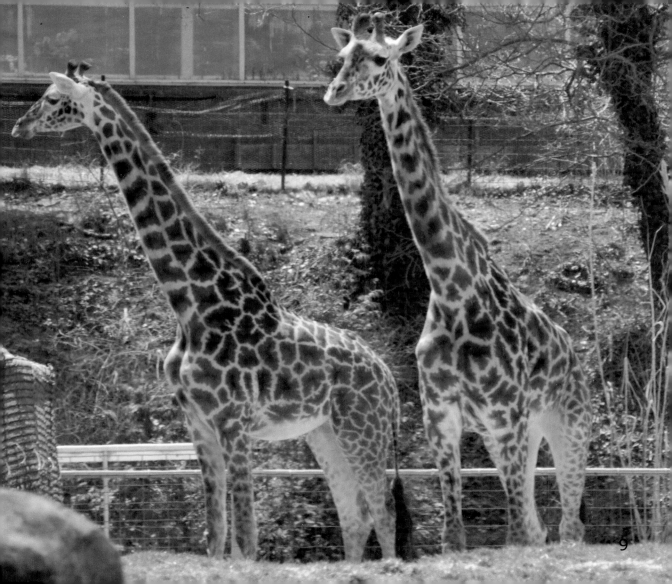

I see an elephant at the zoo.

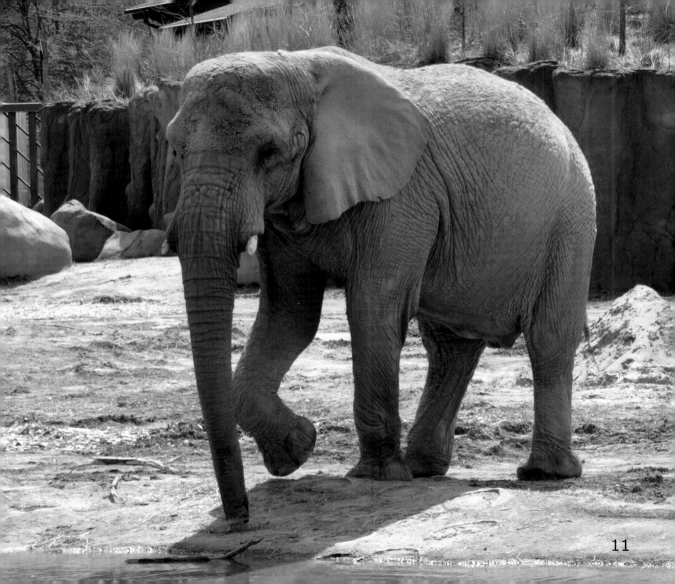

I see camels at the zoo.

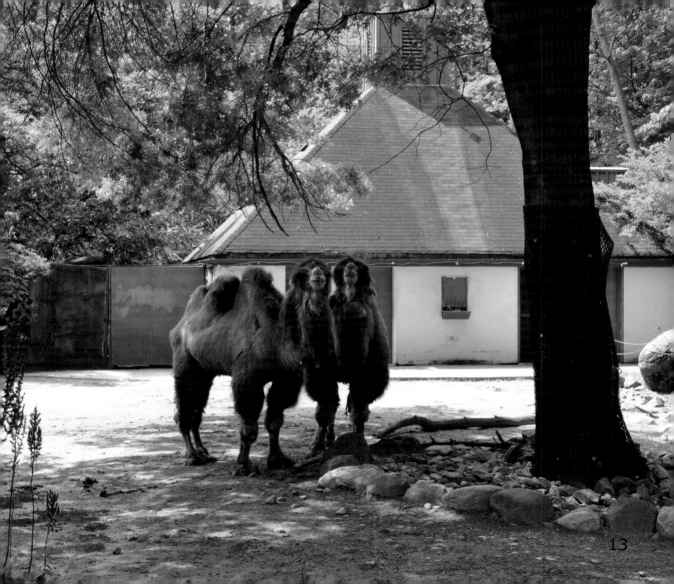

I see a rhinoceros at the zoo.

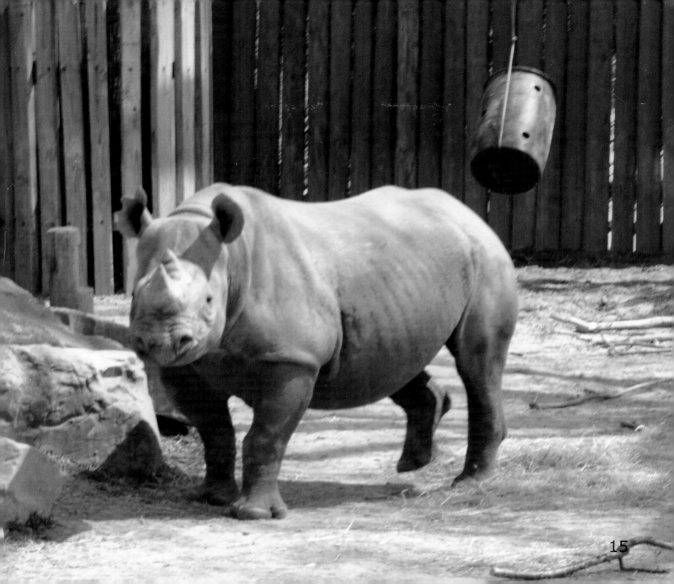

I see a lion at the zoo.

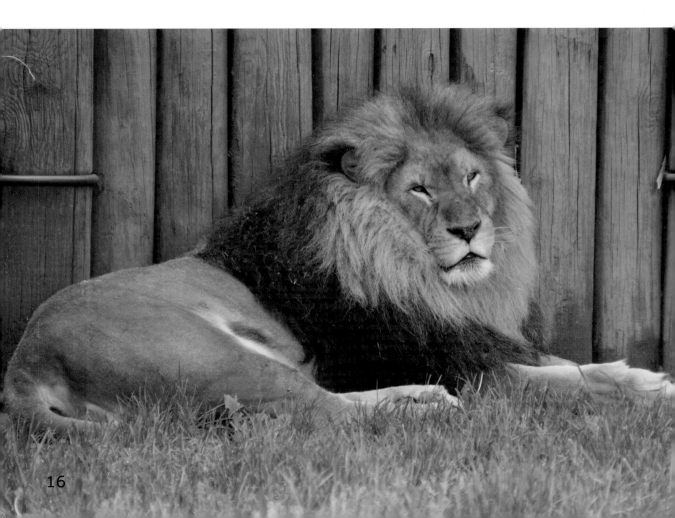